Frontiers in Particle Science and Technology 2020

Held at the 2020 AIChE Spring Meeting and 16th Global Congress on Process Safety

Online
17 – 21 August 2020

ISBN: 978-1-7138-2328-5

Printed from e-media with permission by:

Curran Associates, Inc.
57 Morehouse Lane
Red Hook, NY 12571

Some format issues inherent in the e-media version may also appear in this print version.

Printed with permission by Curran Associates, Inc. (2021)

For permission requests, please contact AIChE
at the address below.

AIChE
120 Wall Street, FL 23
New York, NY 10005-4020

Phone: (800) 242-4363
Fax: (203) 775-5177

www.aiche.org

Additional copies of this publication are available from:

Curran Associates, Inc.
57 Morehouse Lane
Red Hook, NY 12571 USA
Phone: 845-758-0400
Fax: 845-758-2633
Email: curran@proceedings.com
Web: www.proceedings.com

TABLE OF CONTENTS

(195B) DESIGNER PARTICLES: MECHANISTIC UNDERSTANDING FOR PROCESS DESIGN 1
Rachel M. Smith

(196A) SURFACE MODIFICATION OF PARTICLES IN FLUIDISED BEDS AT ATMOSPHERIC CONDITIONS........ 2
Pablo Garcia Triñanes

(203A) USE OF GRANULATION TO DELIVER SILICONE ACTIVES IN POWDERED FORMULATIONS........ 3
Marc Thibaut

(203B) DESIGNER NANOMEDICINES: FORMULATION CONSIDERATIONS FOR THE NEXT GENERATION OF INHALABLE THERAPEUTICS........ 4
Catherine A. Fromen

(205A) TARTRAZINE AND DEXTRAN BLUE AS SURROGATES TO CHARACTERIZE BIOPHARMACEUTICAL PROPERTIES OF COMPACTED MICROCRYSTALLINE CELLULOSE TABLETS........ 5
Aaron Moment

(205B) PHYSICAL CHARACTERIZATION OF BLENDS CONTAINING MESOPOROUS PARTICLES WITH A FOCUS ON ELECTROSTATIC PROPERTIES........ 6
Aurélien Neveu

(208A) PROCESS DESIGN FOR CRYSTAL PROPERTY CONTROL, USING MODEL-BASED OPTIMAL FEEDBACK CONTROL........ 7
Martha A. Grover

(212A) PREPARATION METHODS AND EXAMPLES OF ENGINEERED PARTICLES – SOMETIMES THEY EVEN WORK........ 8
Willie Hendrickson

(212B) RELATING PARTICLE SCALE PROPERTIES TO BULK BEHAVIOR AND THEN TO PROCESS BEHAVIOR (A GUIDE FOR PRODUCT DESIGN: BULLETIN 321 – A PRIMER) 9
Kerry Johanson

(214A) CONTROLLING SEGREGATION BY PARTICLE OR PROCESS DESIGNS THROUGH FUNDAMENTAL UNDERSTANDING 28
Bruce D. Hook

(214B) IMPROVING FLOWABILITY BY DESIGNING POWDERS........ 29
Jayant Khambekar

(216A) MADE TO ORDER: TAILORED FCC CATALYST........ 30
Andishaeh Dadgar

(219A) SPATIALLY-ORIENTED CONTROL IN BATCH CRYSTALLIZATION PROCESSES USING POPULATION BALANCE MODELING........ 31
Marcellus G. F. De Moraes

(219B) PREDICTING THE FLOW OF VARIOUS SIZES AND MOISTURE CONTENTS OF GLASS BEADS 33
Spandana Vajrala

Author Index

613902 Designer Particles: Mechanistic Understanding for Process Design

Abstract:

Particle-liquid and particle-particle interactions are fundamental to a huge range of industrial manufacturing processes, including granulation, coating, drying, and fluidized bed processes. These processes are critical to a wide range of industries, including foods, pharmaceuticals, consumer products, battery manufacture and many more. There exists a considerable, but incomplete, body of understanding of the nano- and micro- scale interactions between particles and particles, and particles and fluids. However, the ability to use this considerable knowledge to predict process performance, to design manufacturing processes, and to design particulate products to give desired performance is generally still considered a research exercise, and industry is often still reliant on substantial experimental trial and error approaches to product and process design.

This talk focuses on two quite different operations: spherical agglomeration, and power coating in tumbling drums. Despite the differences in industrial application and equipment, fundamentally very similar rate processes are occurring at the interfaces between these particles and fluids. In this presentation, both experimental and computational work targeted to identify and quantify mechanisms of these processes will be shown, and the impact of these results on future designer particles and the associated manufacturing processes for these particles will be explored.

614139 Surface Modification of Particles in Fluidised Beds at Atmospheric Conditions

Abstract:

In this study, a method to uniformly functionalize the surface of particles using self-limiting vapor-phase deposition of silanes is presented. The deposition process is carried out at atmospheric pressure in a fluidized bed reactor and the correlation between surface energy and the electrostatic properties of the resulting particulate materials was studied. Glass beads were functionalized with a fluorine-containing silane to change their cohesion properties and therefore inter-particle forces. This demonstrates the potential use of fluidized beds as an alternative route to wet processes producing less wastes with a solvent-free process and achieving more selectivity with shorter processing times. Characterization techniques such as SEM, XPS, TGA, Surface energy and Free energy analysis as well as a method for assessing the electrostatic properties and the polarity of charged glass beads used in this study.

614141 Use of Granulation to Deliver Silicone Actives in Powdered Formulations

Abstract:

Silicones exhibit many useful characteristics, including low thermal conductivity, low chemical reactivity, low toxicity, high thermal stability, resistance to oxygen and UV, high resistance to weathering, low surface tension and capable of wetting most surfaces. Formulating silicone-based material in solid form can often be a challenge as they tend to spread over the surface of any powder. This paper will review some of the approaches that Dow has followed to formulate silicone-based foam control agents used in detergent powders and hydrophobic agents used in the construction industry.

614142 Designer Nanomedicines: Formulation Considerations for the Next Generation of Inhalable Therapeutics

Abstract:

While respiratory diseases globally number among the top causes of mortality, the field of pulmonary drug delivery has lagged other routes of administration in the application of novel therapeutic approaches, in large part due to formulation challenges. Over the past twenty years, advanced knowledge surrounding cellular responses to synthetic nanoparticles has emerged for cancer, tissue regeneration, imaging, and vaccine applications. Such studies have created tremendous opportunity for immune engineering applications, requiring precision designer nanoparticle formulations. Translating this knowledge to inhalable pulmonary drug delivery requires additional design considerations for dry powder stability and aerosolization, as well as delivery of large bio- and macromolecules, coordinated surface functionalization, tunable degradation rates, and precise nanoparticle shape. In this talk, I will discuss emerging trends in the design of nanoparticles for pulmonary drug delivery, as well as specific formulation challenges and considerations for scale-up and translation. Specific to pulmonary drug delivery, I will highlight our recent work addressing the role of surface charge, porosity, particle shape, and degradation rates on immune engineering applications within the lung.

614143 Tartrazine and Dextran Blue as Surrogates to Characterize Biopharmaceutical Properties of Compacted Microcrystalline Cellulose Tablets

Abstract:

Tartrazine and Dextran Blue as Surrogates to Characterize the Biopharmaceutical Properties of Compacted Microcrystalline Cellulose Tablets vis a vis Dissolution and Disintegration in Batch and Flow via UV-VIS Spectrometry

Peng Peng, Gabriella Reis Barham, and Aaron Moment*, *Columbia University*

Biopharmaceutical formulation properties are conveniently assessed indirectly by dissolution and disintegration under controlled conditions. In this work we report using the dyes tartrazine (CAS 1934-21-0) and dextran blue (CAS 87915-38-6, MW 2x106) to assess the dissolution and disintegration rates of microcrystalline cellulose 10 mm tablets made on an MTS Universal Testing Instrument, as a function of compaction pressure, loading of the dye, and test configuration. In batch, the formulations are distinguished using a lumped Noyes-Whitney equation model. On the other hand, with a custom designed flow cell we report more detailed measurements including diffusion coefficients and overall mass transfer coefficients under low Reynolds number conditions mimicking saliva flow for an intra oral delivery project. We also offer a preliminary assessment of image analysis algorithms to assess spatial content uniformity of the formulations based on micrographs and photographs.

Key words: compaction, tensile strength, dissolution, disintegration, dextran blue, tartrazine, FD&C Yellow 5, MTS Universal Testing Instrument, microcrystalline cellulose, image analysis, flow cell, UV-VIS spectroscopy, diffusion coefficient, mass transfer coefficient

614144 Physical characterization of blends containing mesoporous particles with a focus on electrostatic properties

Abstract:

We show the effect of the addition of a small fraction of mesoporous silica (MPS) grains on the flow of cohesive powders. MPS grains are found to decrease the powder cohesiveness by playing a role on both capillary and electrostatic forces. The effect on the flow is quantified with the rotating drum method (GranuDrum instrument). According to both grain and pore size of MPS, the cohesiveness decrease can be immediate (for small MPS grains having small pore size) or on the longer term (for larger MPS grains having higher pore size). The effect on the powder electrostatic properties is analysed with a powder tribo-electrometer called GranuCharge. The quantity of electrostatic charges created during a flow in contact with stainless steel is decreased by the presence of MPS grains. We show that this decrease is induced by a modification of electric resistivity. The MPS grains are adsorbing the humidity contained in the powder and are forming a network having a low resistivity, improving the diffusion of electrostatic charges.

614145 Process Design for Crystal Property Control, Using Model-Based Optimal Feedback Control

Abstract:

Crystal properties such as size, shape, and form are critical to performance in downstream processes and final products. However, it is not straightforward to determine the process design and recipe required to generate target properties reliably. Three barriers are (1) the lack of understanding of the underlying crystallization process, (2) the stochastic nature of crystallization (especially nucleation), and (3) the limited availability of *in situ* sensors that measure crystal properties directly. Here we present an approach to controlling crystal properties using feedback control. The feedback policy is specifically calculated to achieve target properties, using a dynamic model of the process together with *in situ* sensors for feedback. If fundamental physical models are available, they can be used to calculate the policy, but empirical models learned from data can alternatively be used and may provide a more direct path to implementation of crystal property control.

614149 Preparation Methods and Examples of Engineered Particles – Sometimes They Even Work

Abstract:

Rationale for developing and processing particles with specific attributes will be presented. Engineered particles that protect omega-3 oil from oxidation using special microencapsulation methods, modified cellulose fibers with high water holding capacity, and engineered particles produced by melt atomization processes with unique attributes will be discussed. In the examples presented a discussion of alternative processing methods, the importance of understanding the materials being used, and what can happen when you do not understand the functional properties that you are designing for.

212b Relating Particle Scale Properties to Bulk Behavior and Then to Process Behavior (A guide for product design: Bulletin 321 – a primer)

By Kerry Johanson

FPST – AICHE spring Meeting August 17-19, 2020

In the world of powders there exists a complex relationship between the particle scale properties and the behavior that one observes on a plant scale as powders are processed, packaged, or even used by the customer. Over the years the community has developed a means of using bulk properties, through scientific and theory-based procedures and analysis, to describe powder behavior during processing, handling and packaging. This has led to helpful design criteria in designing a powder handling process. It is not completely accurate, still having some limitations, but provides a good set of basic rules to guide engineers in configuring and specifying equipment to handle powders most of the time – if bulk flow properties are known. When it comes to process design, as long as the process is built of simple and basic structures we understand the rules that govern good process design. However, the same cannot be said when establishing how a powder will work in a given handling system if only the basic information about the size, shape and particle surface morphology and other particle based measurements are known. There are two barriers to developing an analysis or design basis relating particle scale properties to process design. The problem is a two scale process. First, one must understand how a change in particle scale will change a set of key bulk scale flow properties. Ten one must understand, select and use a set of bulk scale properties to design the process of interest.

The second step in this process is reasonably well understood. This part of the design process began with some pioneering work in the 1960's and continues today. However, the first step in this process (particle to bulk scale) still requires work so as to understand how basic particle scale properties influence bulk properties. One of the most important bulk properties to understand is the unconfined yield strength. Understanding this property allows design of containers, bin, hoppers, feeders, blenders, fluid beds, milling operations, roll press, and agglomeration units without flow problems. For each unit operation in the aforementioned list there are other key properties that must be understood to create a reliable design. But, unconfined yield strength (cohesion) is common to all these unit operations. Thus, understanding and using the relationship between particle scale properties and cohesive flow properties could be a major step forward in relating particle scale properties to process behavior. The following paper discusses how particle scale properties influence cohesive flow properties.

Introduction

The concept of processes design given a set of flow properties is fairly well understood provided the process is simple, not subject to excessive gas flows, material does not segregate excessively, and flow is symmetric in the process equipment. In the 1960's two researchers Jenike [1],[2] and Johanson [3] embarked on trying to understand how simple flow properties such as unconfined yield strength, wall friction and bulk density might be used to design simple silos, bins, hoppers, and feeders. These design rules are well tested and, when applied, can be used to generate solids handling systems without major flow problems with most materials. The same cannot be said when attempting to design a material to have prescribed flow properties. The unconfined yield strength of a bulk material is the major principle stress that must act on a bulk material in an unconfined state to induce and initial failure or yield by shear (Figure 1). It is used to describe the arching or rathole ability of the material. It can also be used to determine the robustness of lumps that might form in material. This singular property interacts with other flow properties such as bulk density and compressibility. A material that has some unconfined yield strength causes cohesive bonds to form between particles, resulting in a loose packed condition when a bulk material is placed in a container with a small consolidation pressure. However, placing additional consolidation stresses on the material overcomes the bulk unconfined yield strength and collapses the loose packed state, creating a denser material at higher consolidation pressure. In general, a cohesive bulk material will be compressible. However, a free-flowing material will generally be much less compressible. Because the cohesion in the bulk material changes the material's void structure, unconfined yield strength also has an effect on the permeability of the bulk material. Often the cohesive nature of the bulk also causes interaction of the bulk material with wall surfaces. Since, changes in unconfined yield strength can also result in changes in wall friction angles, understanding the root cause behind unconfined yield strength can help determine the flow properties of the bulk material and, hence, the expected behavior in the process or handling system.

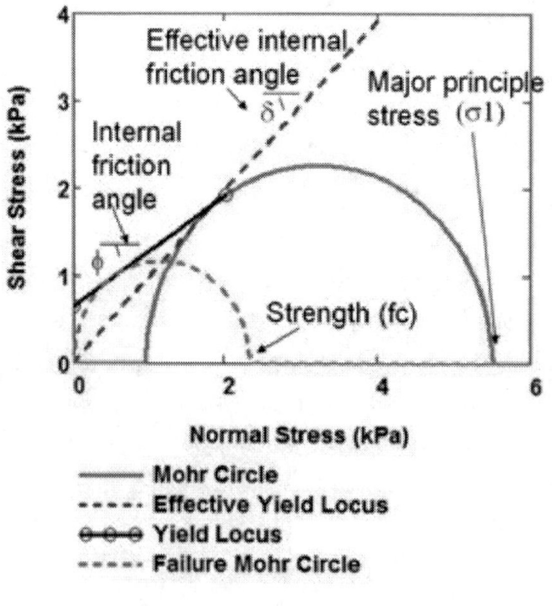

Figure 1. Schematic showing the concept of the bulk unconfined yield strength of material

There are two different ways to look at unconfined yield strength. We can take the traditional soil mechanics approach, defining the cohesion and unconfined yield strength in terms of a continuum method and a constitutive equation describing bulk failure of the material on a yield surface. In this case the strength is written as a function of stresses and internal angles of friction, but not as a function of any particle scale parameters. The continuum approach does not give us a root cause of the bulk strength. Conversely, we can consider the assembly of particles, allowing for the forces acting between particles that might provide a resistance of the relative movement of two or more adjacent particles. As a collection of particles moves relative to each other, adjacent particles slide over one another to induce a set of frictional forces between the particles that adds to the shear resistance (Figure 2). Likewise, two adjacent particles may pull apart (lose contact) during the shearing action of the bulk. In this case, the adhesion force between particles causes a resistance to shear that adds to the bulk unconfined yield strength. Since, the motion of 3D particles relative to each other requires motion of two particles in a direction perpendicular to the plane of movement along the particle surfaces, inducing a confining stress caused by the resistance of motion of all the surrounding particles. When viewed from a particle scale approach, the bulk unconfined yield strength should be a function of the total sum of all forces acting between the shearing particle assembly. Isolating the particles in the act of shearing and then adding up the effect of all the adhesion forces (F_{ad}), particle friction forces (F_s and F_n) and particle confining forces (F and S), yields the stress acting on the bulk material that resists interparticle movement. This approach gives insight regarding the relationship between the type of adhesion, frictional forces, external contact forces, number of particles involved with shear, and the shape of the particles engaged in shear and the bulk unconfined yield strength of the material.

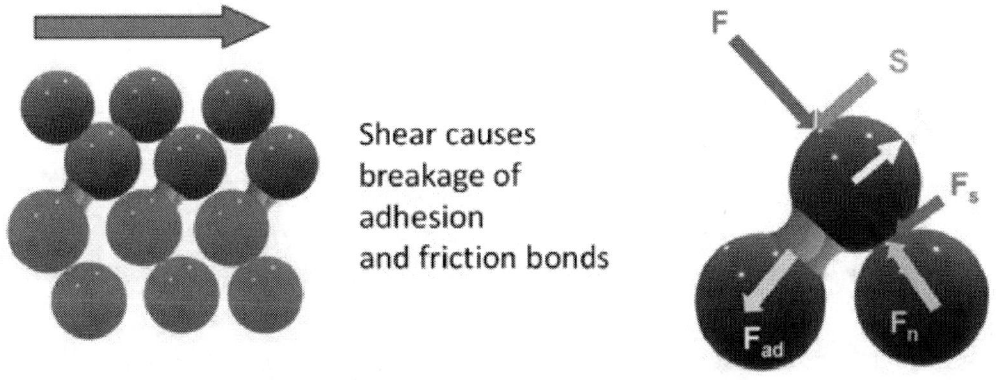

Figure 2. A summary of the forces acting between two adjacent particles during a shearing event

Now consider the work required to move a set of particle (active in shear) one particle's diameter. From a continuum point of view that work (E_{mech}) would be a function of the shear stress acing on the bulk material in the unconfined state (τ) and given by equation 1.

$$E_{mech} = 2 \cdot R \cdot \tau \cdot A_{sl} \qquad (1)$$

One can also add up the work to break individual adhesion bonds or induce motion against frictional contact of adjacent particles (equation 2).

$$W_{step} = W_1 \cdot N \cdot A_{sl} \cdot \frac{n_c}{2} \cdot N_{layer} \qquad (2)$$

In this case, the work to break or move a particle relative to an adjacent particle is a function of the incremental work for a single particle (W_1), the number of particles in the shear layer (n), the area of the shear layer (A_{sl}), the number of shear layer in motion (N_{layer}), and the condonation number of the assembly (n_c). Details of these calculations can be found in the following papers [4], [5], [6] , and [7].

The net effect of this analysis is that, for the case where strength depends on the formation of capillary bonds between adjacent particles or the strength depends on frictional contacts between particles, then the bulk unconfined yield strength is a function of 1/Dp where Dp is the particle diameter (Figure 3).

$$f_c = \frac{8 \times \gamma \times k_r \times n_l \times p \times \cos\theta}{\sqrt[3]{k_h} \times Dp \times \cos\phi} \sqrt{\frac{C \times n_c \times \rho_{sol}}{3 \times \rho_{liq}}} + \frac{\mu \times A_h \times n_c \times n_l \times p \times \sqrt[3]{k_r^{\,2}}}{6 \times \pi \times Dp \times H^2 \times rf \times \cos\phi}$$

γ – surface tension
ρ_{liq} – density of binder
n_c – coordination number
p – probability of rupture
k_h – hexagonal packing factor
rf – roughness factor
ϕ – angle of internal friction
θ – contact angle
μ – friction coefficient
n_l – number of shear layers
k_r – random packing factor
H – separation distance
ρ_{sol} – density of particles
A_h – Hamaker constant

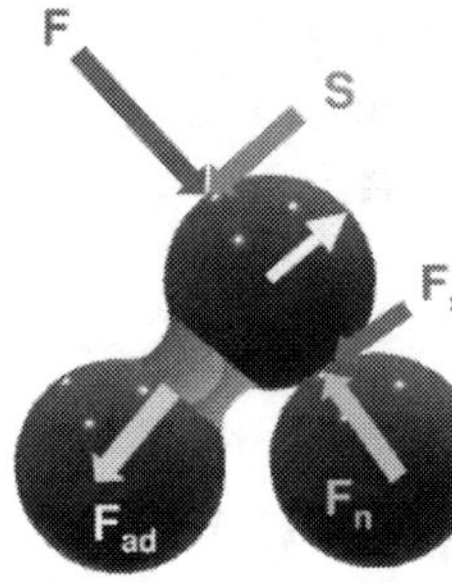

Figure 3. Strength as a function of particle size for capillary and particle friction effects

It is important to note that equation 3 is for the case where the particles are very close in size and only capillary bonds and friction conditions between particles are part of the set of forces that contribute to the bulk unconfined yield strength.

$$f_c = \frac{8 \times \gamma \times k_r \times n_l \times p \times \cos\theta}{\sqrt[3]{k_h} \times Dp \times \cos\phi} \sqrt{\frac{C \times n_c \times \rho_{sol}}{3 \times \rho_{liq}}} + \frac{\mu \times A_h \times n_c \times n_l \times p \times \sqrt[3]{k_r^{\,2}}}{6 \times \pi \times Dp \times H^2 \times rf \times \cos\phi} \qquad (3)$$

To validate this relationship researchers measured the bulk unconfined yield strength using the Schulze method for uniform sized glass beads with oil added to the mixture (Figure 4). The above equation then suggests that the strength is a function of the square root of the moisture content (C), the particle diameter (Dp), and the interparticle friction (μ) (equation 4)

$$f_c = Ka\frac{\sqrt{C}}{Dp} + Kb\frac{\mu}{Dp} \qquad (4)$$

Thus, plotting the unconfined yield strength as a function of the square root of the moisture content should yield a linear line (Figure 5). The intercept on this curve is the effect of inter-particle friction. The slope of this curve is the effect of the particle scale parameters that make up the capillary adhesion force between adjacent particles.

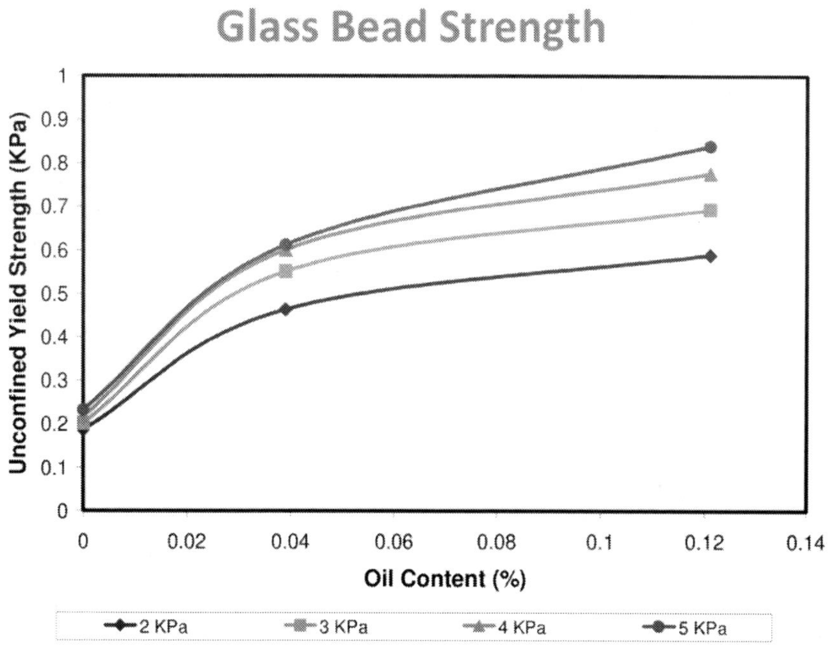

Figure 4. Strength measured as a function of oil content for four different consolidation stress levels

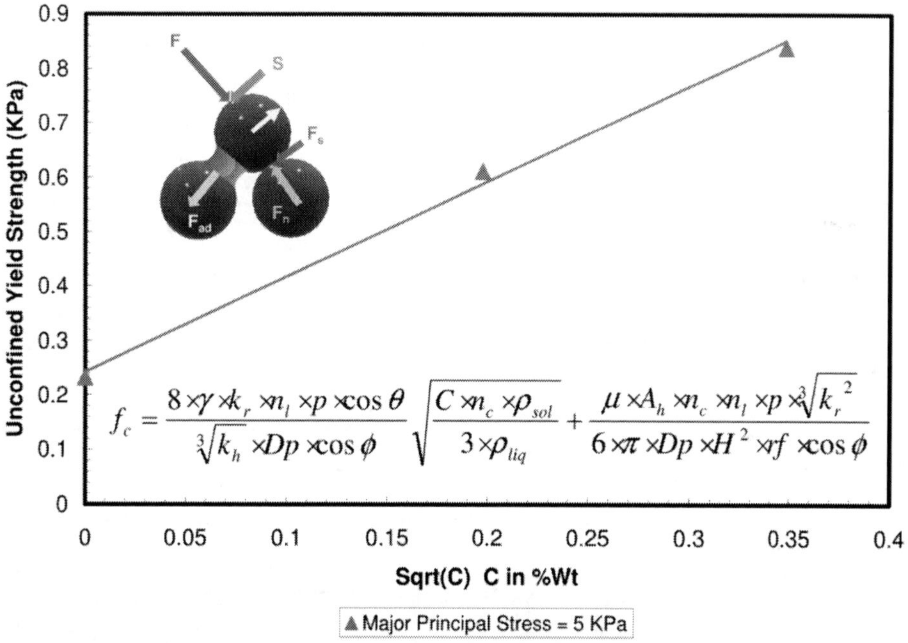

Figure 5. Strength measured as a function of the square root of oil content

There are many things that influence the bulk strength of a powder. Some are related to the forces between particles and some are related to the structure of the bulk powder system (Figure 6). In order to understand how to design a powder with a prescribed strength, both must be understand.

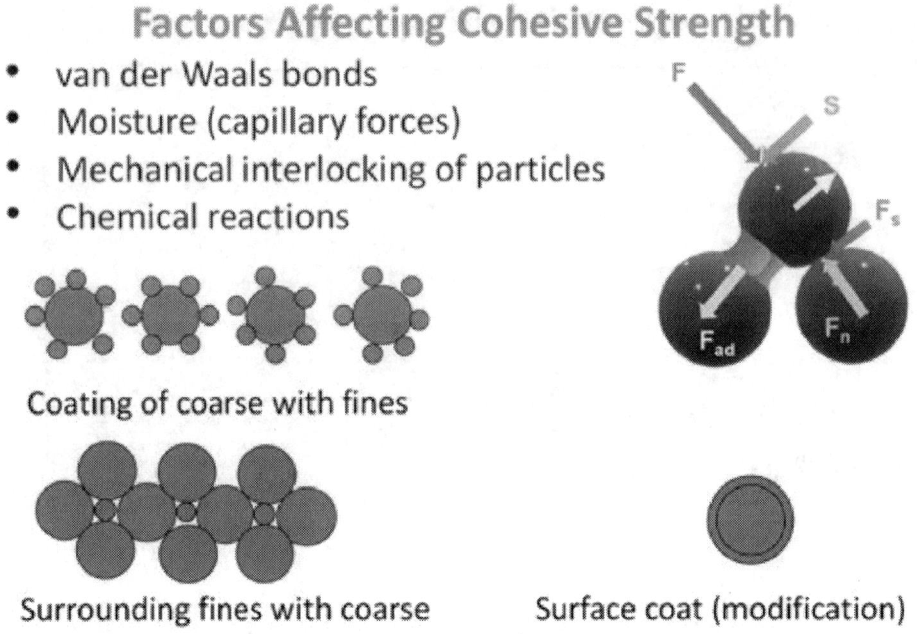

Figure 6. Things that effect bulk unconfined yield strength

Capillary forces are not the only form of adhesive forces that may be present in a bulk powder system. Past researchers have postulated several relationships between bulk unconfined yield strength, particle size and other key particle scale parameters. Some of these are summarized in the Table I below.

Table I: Typical Mechanisms for Bulk Strength

Strength Model for Uniform Particles	Mechanism	Reference
$fc = \frac{K_1}{Dp^2}$	Van der Waals forces	Mollerus [8]
$fc = \frac{K_2 \cdot \sqrt{C}}{Dp}$	Capillary bond formation	Rabinovich [9]
$fc = \frac{K_3 \cdot \mu}{Dp}$	Interparticle friction	Rabinovich [4] , [5]
$fc = \frac{K_4}{Dp^{1/2}}$	Elastic failure of bonds	Rumpf [10]
$fc = \frac{K_5}{Dp^n}$	Plastic elastic failure of bonds	Specht [11]

Each of these models is for the case where the particle size of the material is relatively uniform. However, any bulk solid is made up of many contacts and there is no reason to assume that each contact is uniform with respect to all other contacts. The moisture distribution in a sample is often non-uniform (Figure 7) and depends on the structure and surface characteristics of the powder system.

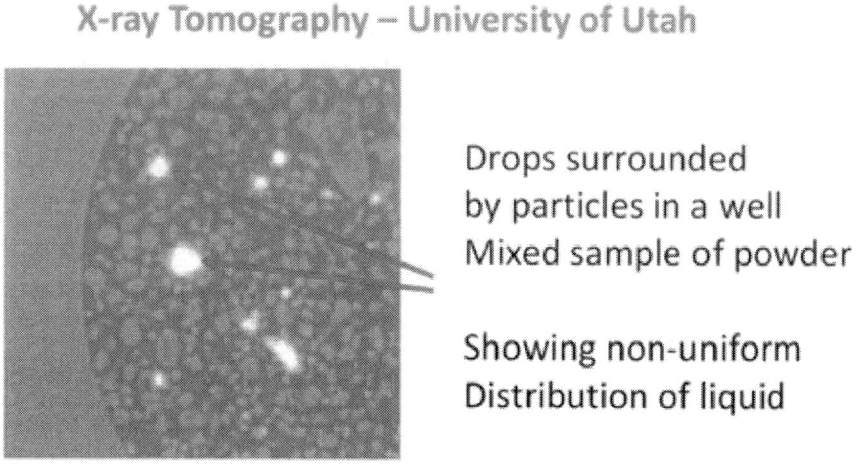

Figure 7. CT Image showing non uniform distribution of liquid in powder sample

Because of this disparity between contacts, the bulk unconfined yield strength can be a function of multiple mechanisms or causes all acting at the same time in a powder sample, depending on the local structure of the material. Thus, the bulk strength is a linear combination of the strengths induced by the various modes or causes outlined in Table I (equation 5).

$$fc = \frac{K_1}{Dp^2} \cdot X_{van} + \frac{K_2 \cdot \sqrt{C}}{Dp} \cdot X_{cap} + \frac{K_3 \cdot \mu}{Dp} \cdot X_{frict} + \frac{K_4}{Dp^{1/2}} \cdot X_e + \frac{K_5}{Dp^n} \cdot X_{ep} \qquad (5)$$

This, then, provides a means of estimating the bulk strength of a given material based on the type of force interactions that cause the strength in the first place. However, this assumes that strength is a linear combination of particle-particle interactions in the bulk. In systems where poly dispersity is a possibility, the linear combination of mechanistic strength laws based simply on particle size may not be enough to characterize bulk unconfined yield strength. Past researchers have included the effect of filling the voids with additional material and found that the maximum strength occurs when all the voids are filled with progressively finer particles. The results are summarized by some work done at the University of Florida [12]. The unconfined yield strength of bimodal mixtures was studied and it was observed that the maximum strength occurred when the percent of fines was between 40% and 55% (Figures 8 and 9).

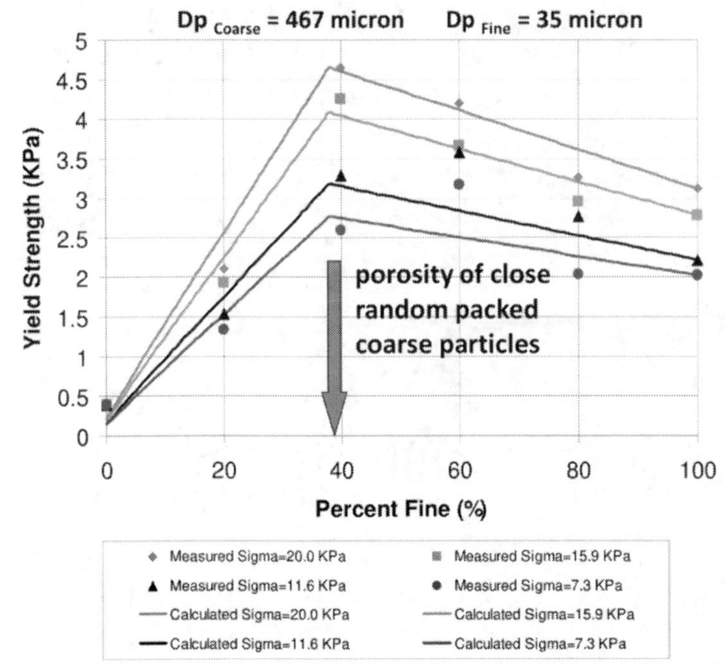

Figure 8. Unconfined yield strength of a mixture of fines and coarse beads; 467 micron coarse, 35 micron fines

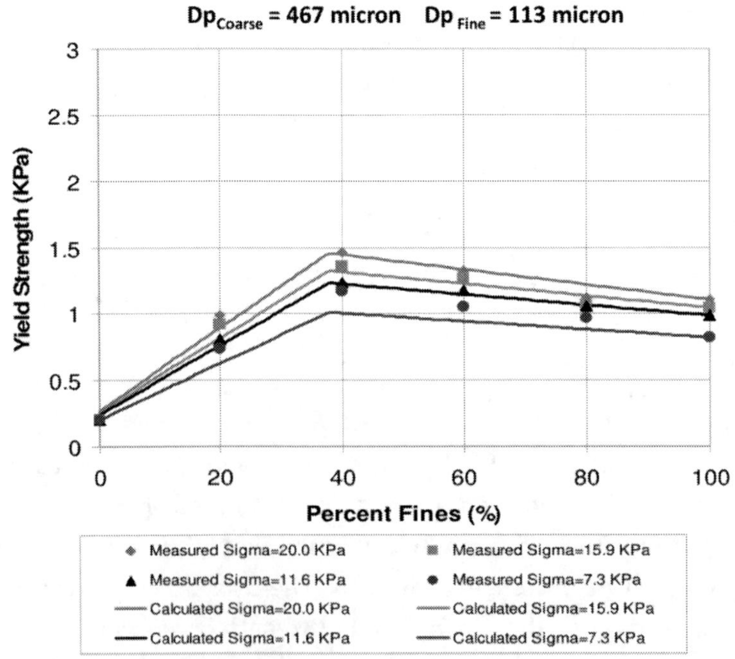

Figure 9. Unconfined yield strength of a mixture of fines and coarse beads; 467 micron coarse, 113 micron fines

It is important to note that the finer material causes a substantial increase in the strength at 40% fines than does the less fine material. Apparently, the more bimodal a material, the stronger the maximum yield strength is. For this reason, non-linear relationship between the bulk unconfined yield strength and the amount of fines in the system, a method of computing the bulk strength of for poly disperse systems was postulated. Consider the following three cases (Figures 10 through 13).

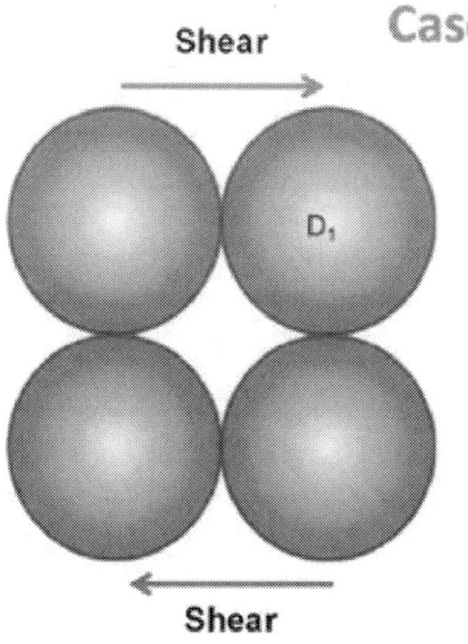

Case 1

There are no small particles in the system. Shear takes place between adjacent large particles. The strength of this structure is found from the strength values for mono-dispersed systems for large particles.

$$fc = \sum_{A} K_1 \cdot \frac{Dp_{50}}{Dp_A} + K_2 \cdot \left(\frac{Dp_{50}}{Dp_A}\right)^2$$

Where Dp_A is determined by the size of the larger particles in the bulk material

Figure 10. ***Unconfined yield strength for the case of all coarse particles in the powder mixture. (with assumed mechanisms for capillary-friction and van der Waals forces)***

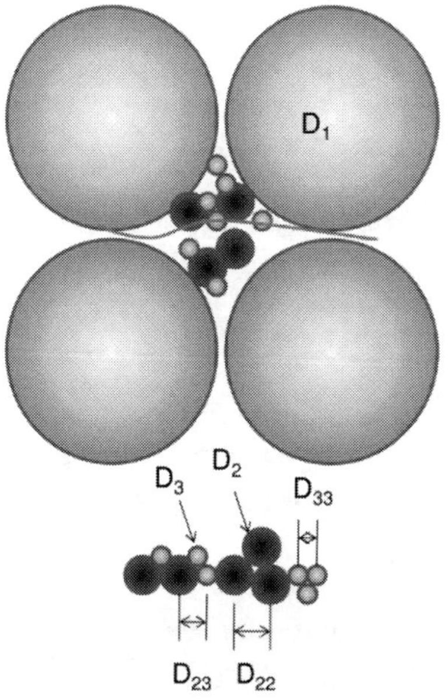

Case 2

There are an excess of small particles in the system. Shear takes place between adjacent small particles. The strength of this structure is found from the strength values for mono-dispersed systems for small particles where Dp_B is the average particle size in the void structure.

$$fc = \left[\left(K_1 \times \frac{Dp_{50}}{Dp_B} + K_2 \times \left(\frac{Dp_{50}}{Dp_B}\right)^2\right)\right]$$

Figure 11. ***Unconfined yield strength for the case of all fine particles in the powder mixture. (with assumed mechanisms for capillary-friction and van der Waals forces)***

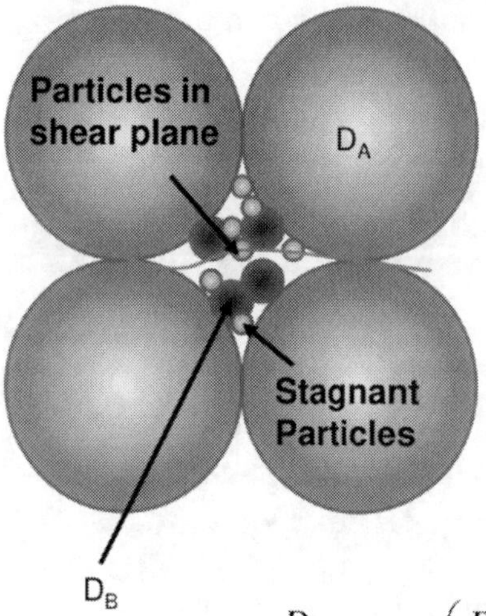

$$fc = \sum_A K_1 \times \frac{Dp_{50}}{Dp_A} + K_2 \times \left(\frac{Dp_{50}}{Dp_A}\right)^2 + \left[\left(K_1 \times \frac{Dp_{50}}{Dp_B} + K_2 \times \left(\frac{Dp_{50}}{Dp_B}\right)^2\right) \times n_{shear}\right]$$

Figure 12. Unconfined yield strength for the case of coarse particles with fines filling the voids in the powder mixture (with assumed mechanisms for capillary-friction and van der Waals forces)

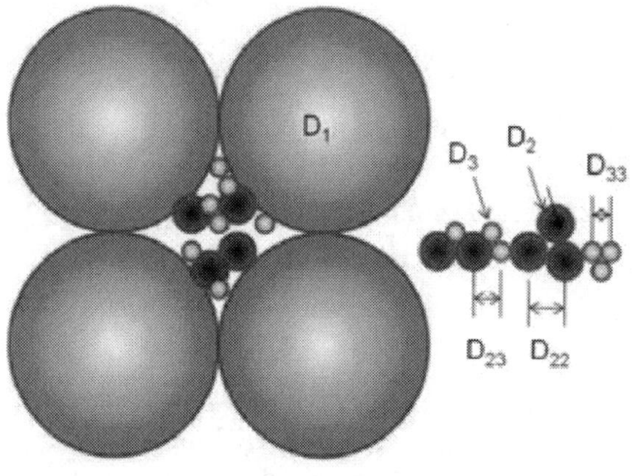

$$D_{avg} = D_{22} \cdot (f_2^3 + f_2^2 \cdot f_3) + 2 \cdot D_{22} \cdot (f_3^2 \cdot f_2 + f_2^2 \cdot f_3) + D_{33} \cdot (f_3^3 + f_3^2 \cdot f_2)$$

f_2 is the fraction of #2 particles in the pores
f_3 is the fraction of #3 particles in the pores

Figure 13. Unconfined yield strength for the case of coarse particles with a mixture of fines and coarse filling the voids in the powder mixture (with assumed mechanisms for capillary-friction and van der Waals forces)

The technique described above was applied to three samples of ibuprofen, each having a different particle size distribution. The strength was measured using the SSSpinTester® instrument. The SSSPinTester® is a means of measuring the unconfined yield strength using a small amount of material and able to estimate the bulk unconfined yield strength at very low consolidation stress levels. The procedure uses centrifugal force for consolidating and failing the material. The basic steps of this measurement are as follows:

Step 1:

A thick guard is inserted below the larger diameter opening of the test cell's conical orifice and material is carefully introduced into the cell through the outlet by passing/vibrating powder through a coarse sieve to break agglomerates. The guard mates with the conical cell cavity. This combination constitutes a small bin which is filled with powder. The gentle fill process reduces over-compaction pressures that arise during filling, allowing strength to be measured at very low consolidation pressures. Over-consolidation due to handling is reduced. This is not as critical when measuring at large consolidation pressures. The weight of the material in the cell after filling is recorded and is used to compute the consolidation pressure (Figure 14).

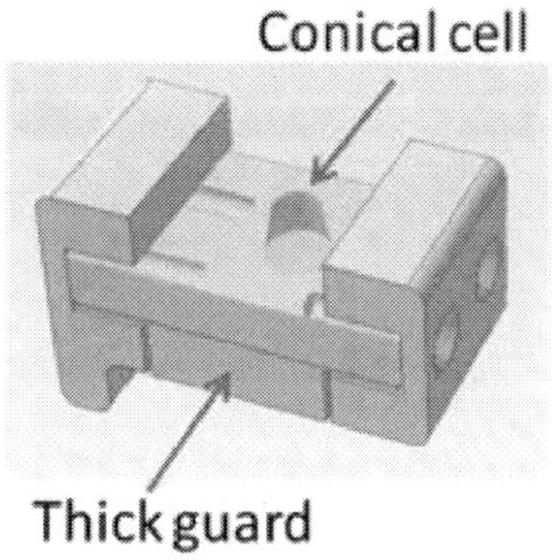

Figure 14. Schematic of step 1

Step 2:

The second step involves placing a flat guard at the top of the cell and then setting the cell in a rotary cavity such that the axis of the conical cavity is 90 degrees from the direction of rotation and at a prescribed distance from the axis of rotation. Care should be taken to place this cell in the rotor without jarring or sudden acceleration that might compact the material (Figure 15).

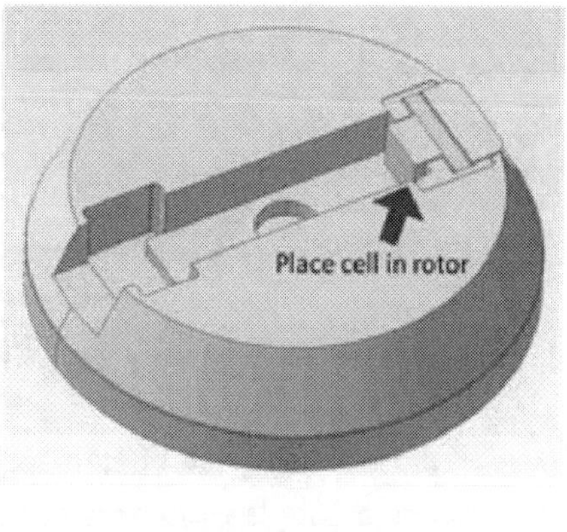

Figure 15. Schematic of step 2

Step 3:

The cell and rotor are rotated to a prescribed speed and that prescribed speed is held for an allotted time, causing centrifugal forces to act on the bulk material in the conical cell and compact the material. The rotor speed and weight of the material, with the position relative to the axis of rotation, are used to compute the consolidation pressure. After consolidation, the cell is removed from the rotor and the thick guard is pulled across the conical cell, scraping away excess material and leaving the entire thickness of the cell full. A thin guard is temporarily inserted in its place. The weight of material in just the conical section is recorded and will be used to compute the yield strength (Figure 16).

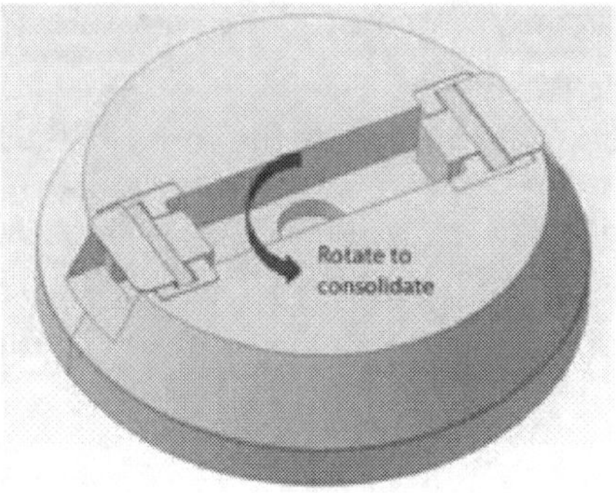

Figure 16. Schematic of step 3

Step 4:

The cell replaced in the rotor as shown. Both guards are carefully removed. The bulk material presumably has strength and will arch over the conical cavity. Rotation speed is increased incrementally until the compacted material exits the conical cavity due to centrifugal force. The weight of the material, the position relative to the axis of rotation, and the rotation speed at the point when the material leaves the cell are used to compute the force needed to fail the compacted material in the conical arch. This data is used to compute the bulk unconfined yield strength (Figure 17).

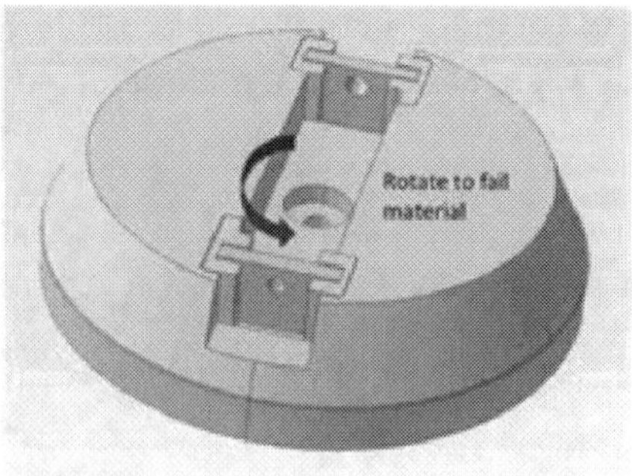

Figure 17. Schematic of step 4

The full process requires about 0.1 cc of material and can be completed in just a few minutes. The user interface requires filling a cell and removing a couple of guards at key points during the test, as well as acquiring and recording to the machine the weight of the material sample. The test technique is accomplished in 3 to 5 minutes for each strength measurement of interest. The results are reported in Figure 18.

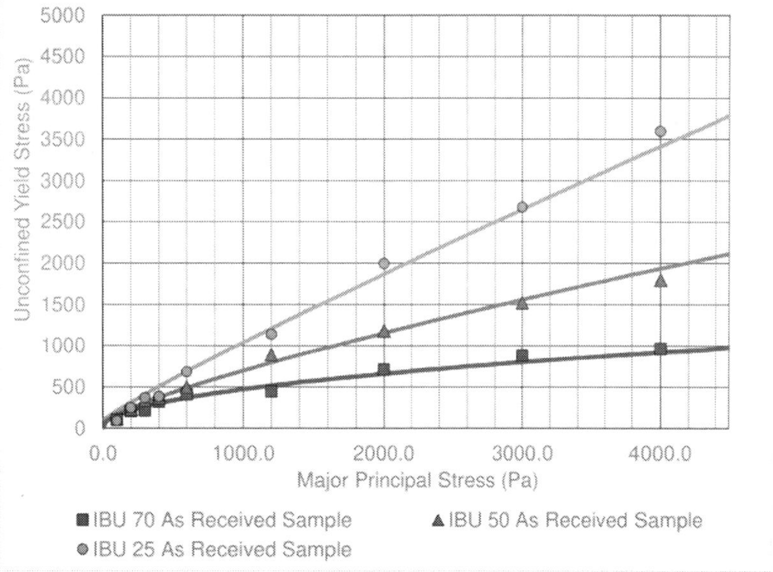

Figure 18. Unconfined yield strength of Ibuprofen samples as a function of major principal stress for three samples of ibuprofen

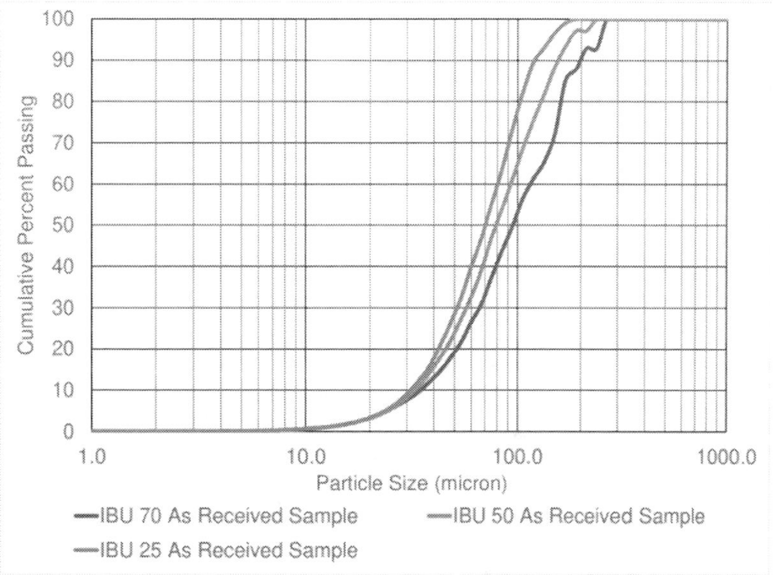

Figure 19. Particle size distribution based on volume for three samples of Ibuprofen

The particle size of these three sample of ibuprofen material was also measured using an optical technique (Figure 19). The particle size distributions are presented as cumulative distributions based on volume and expressed as percent passing. The samples were also sieved to obtain various sieve cuts and the strength of these sieve cuts were measured using the SSSpinTester®. The data was plotted for each consolidation stress level as a function of the average sieve size particle diameter (Figure 20). Each plot showed an increase in the strength at smaller particle size distributions.

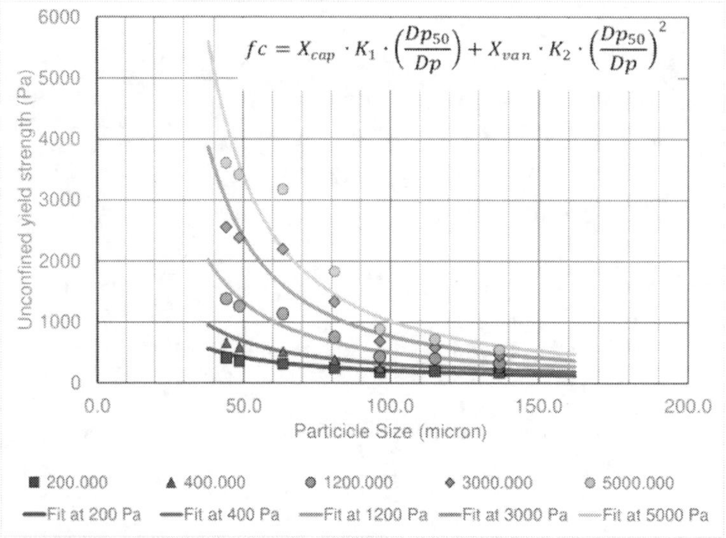

Figure 20. Strength of sieved sizes of Ibuprofen plotted as a function of particle size

It was assumed that the active mechanisms in this powder system were capillary-friction and van der Waals forces, and the values for K_1, K_2, X_{van} and X_{cap} were computed for each particle size sieve cut. The data suggest that 55% of the forces in the powder are due to capillary-friction forces while 45% are due to van der Waals forces. The data also suggest that the K_1 and K_2 values are functions of the consolidation stress level (Figure 21).

$$fc = X_{cap} \cdot K_1 \cdot \left(\frac{Dp_{50}}{Dp}\right) + X_{van} \cdot K_2 \cdot \left(\frac{Dp_{50}}{Dp}\right)^2$$

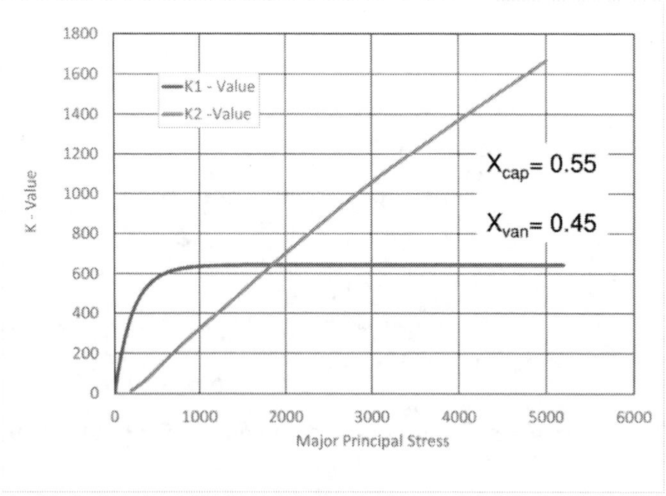

Figure 21. K1 and K2 values as a function of consolidation stress

Note that the K1 value tends to plateau at higher consolidation stresses while the K2 values tends to increase with increasing consolidation stress level. This may be due to the increase in contact area between particles at higher consolidation stress values.

The final task is to divide the particle size distributions for the three ibuprofen sample cuts based on the distribution of particles above and below the D_{50} particle size of the full-size particle size distribution.

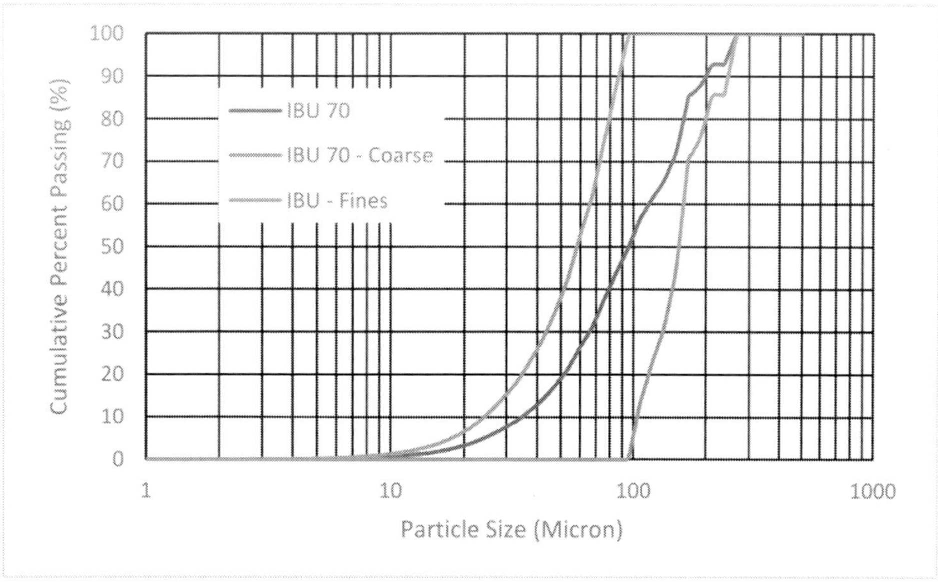

Figure 22. Particle size distribution division for IBU 70 material

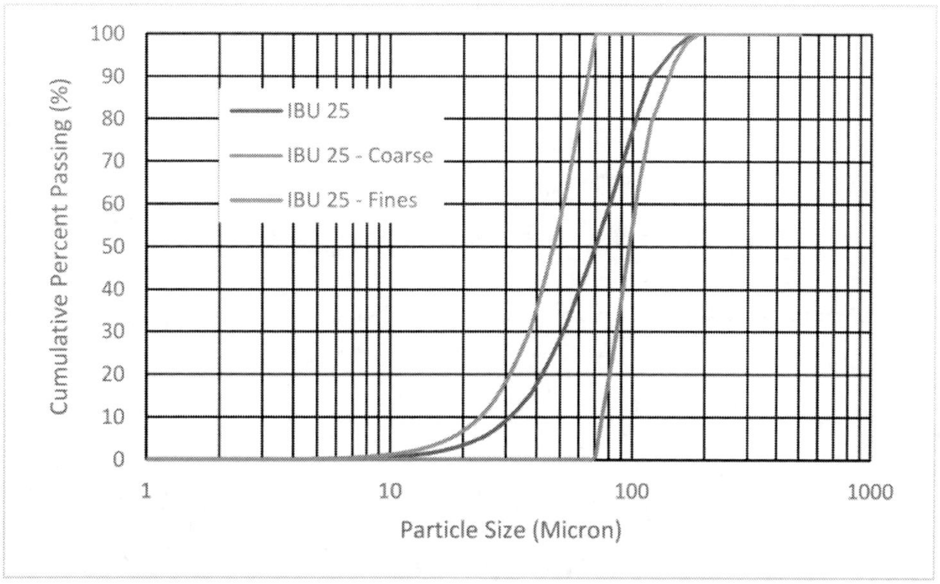

Figure 23. Particle size distribution division for IBU 25 material

The final step in the procedure is to select a randomly particle size from the coarse particle size distribution (orange curve) and create a cubic unit cell with a particle having the randomly selected particle diameter placed at each of the eight corners of the cell. The volume of the void space between these eight particles is then computed. Finally, the average particle size D_{50} for the fines fraction of the IBU (ibuprofen) samples (gray curve) was computed and used to compute the number of particles that would fit in the void space between the eight particles. The strength was computed using the computed K_1 and K_2 values, and the equations for each powder fill case outlined in Figures 10 through 13. This procedure was repeated 100 times with a new selected random particle size from the coarse particle size fraction from Figure 22 and 23. The strength for all these cases were averaged to yield the expected strength for the full mixture of ibuprofen. These computed strength values were then compared against the measured strength values for each IBU sample (Figures 24 through 26). Values for the IBU 25 and IBU 50 samples agree very well, while values for the IBU 70 sample predict a slightly greater strength than measured. This indicates that the methodology outlined in this paper can be used to predict the strength of real powder systems.

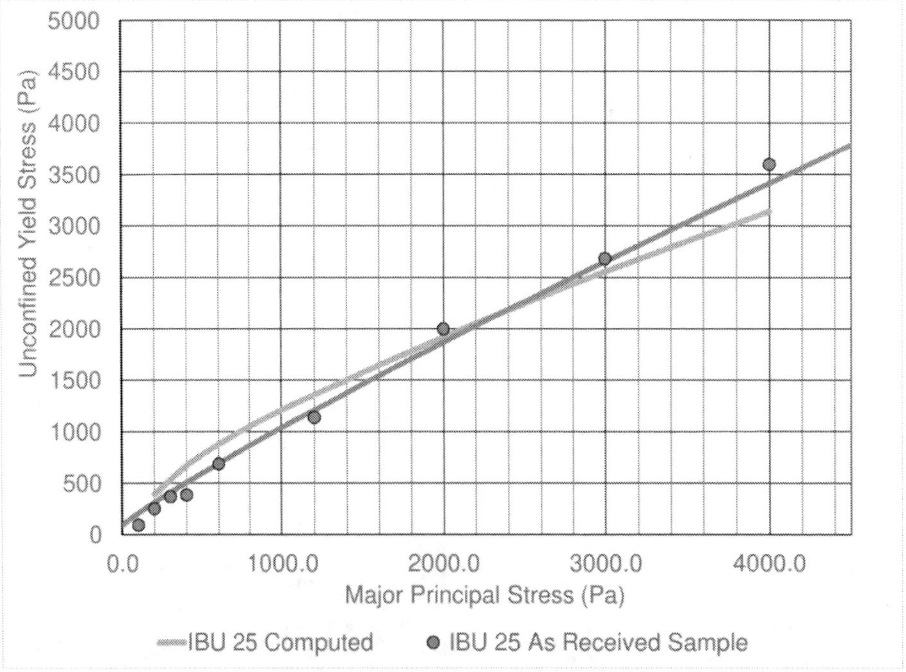

Figure 24. Comparison of strength computed from polydisperse analysis and the measured values for IBU 25

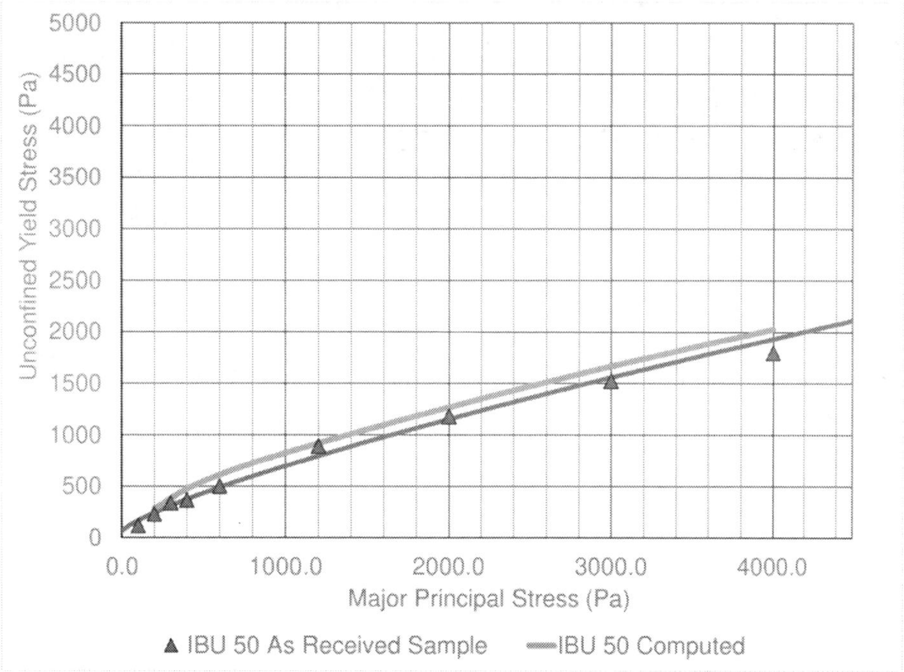

***Figure 25.* Comparison of strength computed from polydisperse analysis and the measured values for IBU 50**

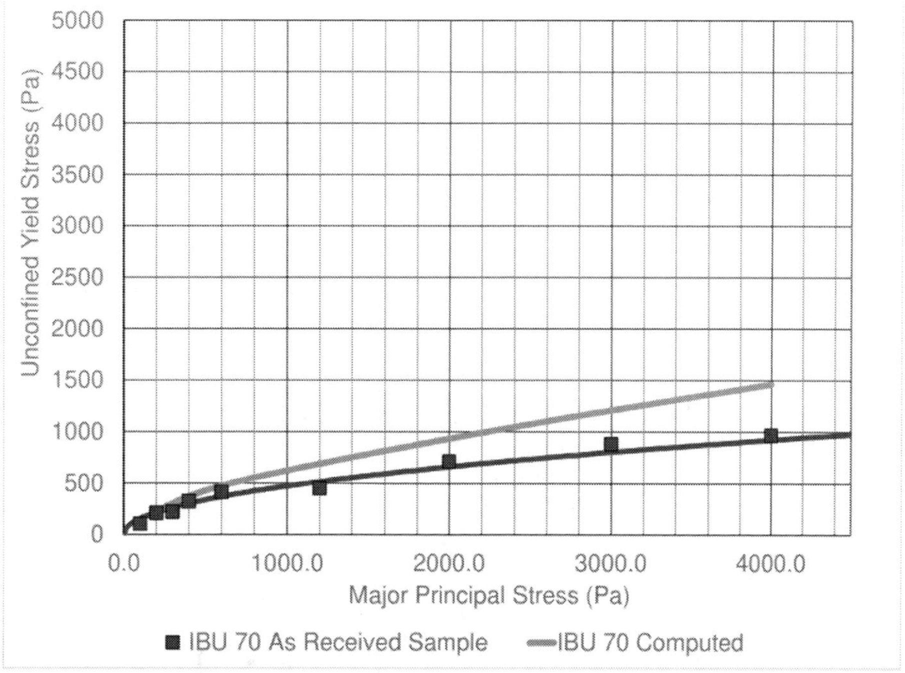

***Figure 26.* Comparison of strength computed from polydisperse analysis and the measured values for IBU 75**

Conclusions

This paper presents a summary of work done with various researchers from the University of Florida and University of Utah in developing the basic equations to begin to understand how to change the particle size distribution to create a prescribed unconfined yield strength. The basis of this method has its root in understanding the relationship between the bulk unconfined strength and particle size from first principles. The relationship between the bulk strength and particle size depends on the cause of the strength. This determines the power law that governs the strength and particle size relationship. A given material can be divided into particle size bins and the bulk strength measured. This strength data can then be used to determine the power laws that dictate the strength behavior. Then, poly disperse models can be used to combine the power laws to predict the bulk unconfined strength for any particle size distribution desired, thus, giving the engineer the ability to design a given material to generate a prescribed strength.

References

[1] A.W. Jenike, Gravity Flow of Bulk Solids, Bulletin, vol. 108, Utah Engineering Station, 1961.

[2] A.W. Jenike, Flow and Storage of Bulk Solids, Bulletin, vol. 123, Utah Engineering Station, 1967.

[3] A.W. Jenike, J.R. Johanson, Stress and Velocity Fields in Gravity Flow of Bulk Solids, Bulletin, vol. 116, Utah Engineering Station, 1967.

[4] K Johanson, Y Rabinovich, B Moudgil, K Breece, H Taylor, Relationship between particle scale capillary forces and bulk unconfined yield strength, Powder Technology, 138 (2003) 13– 17

[5] Y Rabinovich, M Esayanur, K Johanson, C Lin , J Miller, B. M. Moudgil, The flow behavior of the liquid/powder mixture, theory and experiment. I. The effect of the capillary force (bridging rupture), Powder Technology 204 (2010) 173–179

[6] Y Rabinovich, K Johanson, B Moudgil, The flow behavior of the liquid/powder mixture, theory and experiment II: Effect of the friction force, Powder Technology 204 (2010) 180–187

[7] K Johanson, J.D. Miller, C.L. Lin, The effect of hydrophobic surface modification on bulk cohesive strength, Powder Technology 194 (2009) 1–9

[8] O. Molerus, The role of science in particle technology, Powder Technology 122 (2-3, 22) (January 2002) 156–167.

[9] Y. Rabinovich, M.S. Esayanur, K. Johanson, B. Moudgil, Oil mediated particulate adhesion and mechanical properties of powder, Proceedings World Congress on Particle Technology: paper 370, 2002.

[10] H. Rumpf, Particle Technology, Chapman and Hall, New York, 1990.

[11] D. Specht, Caking of granular materials: an experimental and theoretical study, Thesis, University of Florida, 2006.

[12] M.S. Esayanur , INTERPARTICLE FORCE BASED METHODOLOGY FOR PREDICTION OF COHESIVE POWDER FLOW PROPERTIES, Thesis, University of Florida, 2005

614151 Controlling Segregation by Particle or Process Designs through Fundamental Understanding

Abstract:

Particulate products tend to segregate due to size differences of different components when handled, processed, or transported in a variety of industries ranging from chemical to pharmaceutical to agricultural. Segregation can cause unwanted product quality issues or process upset, resulting in significant financial losses due to the need of remixing these products or off specs products. There are many different mechanisms for particle size segregation including sifting/percolation, angle of repose, bouncing, and fluidization, where the sifting and percolation mechanism is the most prominent driving mechanism for granular materials above 200 microns. This talk presents an approach to control segregation of granular materials by properly designing particles (e.g. size or shape) or process conditions based on a recently developed continuum model [1], where particle properties and process conditions are incorporated. A few examples are presented to show how particle size, size distribution, and shape can be modified or process conditions such as feed flow rates can be regulated to minimize segregation. In a rare case, we also show an example of how the approach can be used to optimize process conditions to utilize segregation to achieve better process efficiency and product quality.

[1], Y. Fan, C. Schlick, P. Umbanhowar, J. Ottino, and Lueptow, "Modelling size segregation of granular materials: the roles of segregation, advection and diffusion", Journal of Fluid Mechanics 741, 252-279, 2019.

614152 Improving Flowability by Designing Powders

Abstract:

In many industrial applications, storage and flow of powdered materials are often encountered. Powder flowability is a function of the powder itself as well as of design of the equipment used to handle it. These are the two sides of the same coin; you can achieve better flow behavior by using either approach. Thus, it is quite possible to design powder characteristics in order to achieve good flow behavior. This presentation will discuss various powder characteristics that play a role in determining flow behavior, and also demonstrate by giving examples from industrial applications.

614153 Made to Order: Tailored FCC Catalyst

Abstract:

The main goal of the Fluidized Catalytic Cracking (FCC) unit is to produce transportation fuels and chemicals feedstock from heavy low value petroleum fractions. As oil fields change, the feed to the FCC unit also changes. Some of the new fields are bringing heavier and more contaminated oil to the FCC units. This necessitates the improvement of FCC catalysts to meet demands for better activity and selectivity when processing these heavier more contaminated feedstocks. To accomplish this objective, new materials must be incorporated into the FCC catalyst to boost accessibility, activity, and tolerance. Particle properties need to be carefully balanced to maintain attrition resistance, particle size distribution, and bulk density. Albemarle will show how proper selection of drying conditions and binding systems can be implemented to manufacture catalysts with improved activity and no deterioration of physical properties. Furthermore, catalyst accessibility, which plays a critical role in catalyst performance, will be examined. It is necessary that this physical property is taken into account during the particle formation process for the FCC catalyst.

614718 Spatially-Oriented Control in Batch Crystallization Processes Using Population Balance Modeling

Abstract:

Crystallization is an important separation and purification process with relevant applications in the food, pharmaceutical, fine chemical and fertilizing industries. In these processes, modeling and process control for crystallization are focused on crystal size distribution (CSD) and shape of the crystals in order to make it possible to achieve the desired quality and purity of the product of interest. The control of crystal size and shape is of considerable importance for industrial crystallization processes, as crystal morphology has great influence on downstream steps and also from the point of view of product specification criteria. The consideration of crystal shape in modeling of crystallization processes is also important due to its impact on product effectiveness, such as the bioavailability and tablet stability in pharmaceuticals. Control of the crystallization process is can be facilitated with online measurements, which provide information to be used in the modeling of the process. However, if online measurements are not available, the strongly nonlinear dynamics of these processes makes difficult to establish relations between some inputs (e.g. temperature) and the final desirable characteristics of the crystal product. In this context, image analysis has been shown to be a suitable technique to monitoring the process [1].

In our present work, a population balance model-based scheme for controlling the size and shape of crystals is presented. The modeling was performed using a two-dimensional population balance (2D-PBM) in terms of the two characteristic dimensions of the crystals, describing the evolution of crystal size and shape distributions along the batch time. To identify the nucleation, growth and dissolution kinetics parameters using the 2D-PBM approach along the length and width directions, QICPIC-LIXELL's high speed dynamic image analysis equipment was used for the online measurement of the CSD, shape parameters and aspect ratio distribution of crystals during the experiments. In this way, the developed model allowed for the prediction of the crystal size and shape distributions during growth-dissolution cycles.

The developed framework was employed to obtain optimal temperature policies for controlling the crystal size and shape distributions that meet target criteria for size, morphology and productivity using the dynamic image analysis along the runs. Based on the concept of using measurements along batch crystallization processes as a trajectory in a phase space [2], a trajectory-endpoint control problem is presented with an oriented control scheme that is applied to the shape of the crystals. The scheme was used to control the batch cooling crystallization of potassium dihydrogen phosphate. The results demonstrate the application of the control policies for temperature to produce crystals of the desired average shape and size in different batch times. Thus, using the first principle modeling coupled to the developed trajectory tracking control, the proposed framework has the potential to be applied to industrial crystallization processes for shape control and to be extended to polymorph control investigations.

References:

[1] Eisenschmidt, H.; Bajcinca, N.; Sundmacher, K. (2016) Optimal Control of Crystal Shapes in Batch Crystallization Experiments by Growth-Dissolution Cycles. Crystal Growth & Design, v. 16, p. 3297-2803.

[2] Griffin, D. J., Grover, M. A., Kawajiri, Y., Rousseau, R. W. (2015) Mass-count plots for crystal size control. Chemical Engineering Science, v. 137, p. 338-351.

614717 Predicting The Flow of Various Sizes and Moisture Contents of Glass Beads

Abstract:

Whether it is pneumatic conveying, die filling in a tablet press, or storage in hoppers, powders will be subjected to some level of stress consolidation during their handling, affecting the flow of these materials. Flow issues are likely to occur over a wide range of particle-scale parameters like particle size, particle size distribution (PSD), particle shape and moisture level. For effective product development and process design, it is important to study the flowability of granular materials and develop scale-up equations which can be used on an industrial scale. The present work investigated the effects of particle sizes between 5 – 600 microns and moisture levels from 0 – 20% on the flowability and shear properties of silica beads using the Freeman FT4 Rheometer. A mathematical model in terms of a flowability descriptor called the flow function (FF) was developed, based on the material's yield point. The yield point for various particle size distributions and moisture levels was evaluated using a shear cell test. The FF for dry glass beads increased with particle size, following a linear relationship, as expected. On the other hand, the flow function for wet glass beads followed a power law relationship, where FF decreased with increasing moisture content. However, as the moisture level reached 20%, the flow function value escalated again due to supersaturation. The linear and power functions were combined to develop an empirical relation, which can be used to predict the flowability of glass beads, based on the particle size and moisture levels up to 20%.

AUTHOR INDEX

Dadgar, Andishaeh 30
De Moraes, Marcellus G. F. 31
Fromen, Catherine A. 4
Grover, Martha A. 7
Hendrickson, Willie 8
Hook, Bruce D. 28
Johanson, Kerry 9
Khambekar, Jayant 29
Moment, Aaron 5
Neveu, Aurélien 6
Smith, Rachel M. 1
Thibaut, Marc 3
Triñanes, Pablo Garcia 2
Vajrala, Spandana 33